Surplus gas
Production platform
Drilling derrick
Telecommunications
THISTLE
211/18

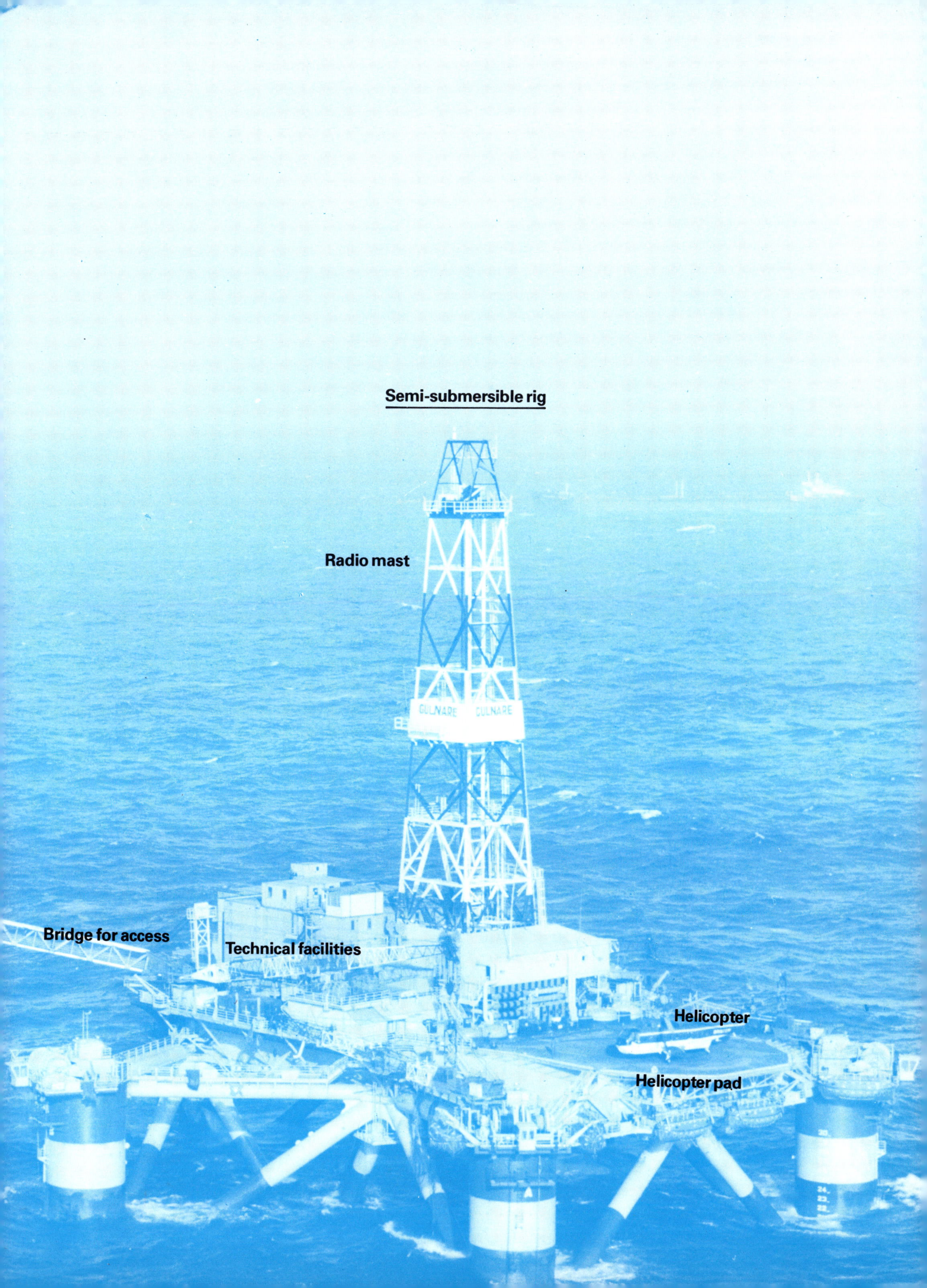
Semi-submersible rig
Radio mast
GULNARE
GULNARE
Bridge for access
Technical facilities
Helicopter
Helicopter pad

First published in 1980 by
Macmillan Children's Books
a division of Macmillan Publishers Limited,
4 Little Essex Street, London WC2R 3LF
and Basingstoke

Published in the United States
by Silver Burdett Company,
Morristown, N.J.
1981 Printing

ISBN 0-382-06616-2

Library of Congress
Catalog No. 81-51497

Designer
Julian Holland
Picture researcher
Tessa Pollitzer
Illustrator
Tony Gibbons/Linden Artists

Photo credits:
British Aerospace; Chris Barker;
J. Allan Cash;
Central Office of Information, London;
Mary Evans; Laurence Gould;
Institute of Oceanographic Sciences; Lockheed;
Magnum; National Maritime Museum;
Natural History Photographic Agency;
Naval Photo Center; NOSC; Photri;
Popperfoto; City of Portsmouth;
The Science Museum, London; Scripps;
Seaphot; L. & M. Taylor Ltd;
Vickers Oceanics; Zefa.
Cover: Seaphot.

Exploration and Discovery

The Undersea World

Series Editor: John Becklake

Author: Thomas Wright

Contents

Left: This diver is using a 'diver-propulsion' vehicle. It is like an underwater moped and the diver uses it to move around while he is working.

Why look under the sea?

The sea covers a large part of the surface of the earth. It also affects our weather and allows life to exist on our planet. Because of this, it is very important to look under the surface of the sea to discover all its hidden secrets. The science of studying the sea is called oceanography.

When earth is viewed from space, the huge area that the sea covers is seen very clearly. Over seven-tenths of the earth's surface is sea. That is why it is often called the blue or watery planet.

Above: *H.M.S. Challenger* was the first ship to go on a voyage to study the sea. It sailed around the world between 1872 and 1876.

Below: A view of the northern half of the globe. You can see the vast oceans which separate the continents.

Above: Rockets can be hidden and launched in the sea.

Above: This diver is exploring a coral reef. He has weights on his belt to help him stay under water.

Above: This rig is tapping oil from beneath the seabed. The sea is now an important energy source.

One of the first things that led man to explore under water was his search for treasure buried in sunken ships. In the old days this was very difficult because the divers did not have equipment to help them stay under water for any length of time. Today, undersea exploration is much easier. Divers use new machines which can go down deeper and stay down longer than ever before. These machines now help us in our search for oil and minerals which are found beneath the seabed. They can help us make maps of the ocean floor, too.

The sea is also studied because of the fish and animals that live in it. As the world population is growing very fast, the sea is an important source of extra food. If we learn about the habits of fish and other sea creatures, we will be able to catch them more easily and efficiently. We will also be able to preserve any threatened species.

Undersea exploration is useful in many ways but it can often be simply enjoyable. Many people today take up diving as a hobby. They do not go very deep but even so they can catch a fascinating glimpse of the very different world that exists under the oceans.

The nature of the challenge

Man meets many problems when he explores the sea. The depth of the sea can vary from 0 to 11,000 metres. As water is much heavier than air, the deeper one goes, the greater the pressure of the water is. Light only reaches a certain depth. The further down one goes, the darker it becomes. In some parts of the sea it is so deep that there is no light at all. Then, as the sea becomes darker, it also becomes colder. The deepest waters are nearly freezing.

These conditions of high pressure, low temperature and little light, as well as the risk of drowning, are the chief difficulties that man has to overcome when he wishes to explore the ocean. Because of these problems, submarines and submersibles are used for very deep diving. They provide protection and allow man to stay under water for long periods of time.

Right: The deeper the sea is, the higher the pressure and the lower the temperature and light become. High pressure has a dangerous effect on the human body.

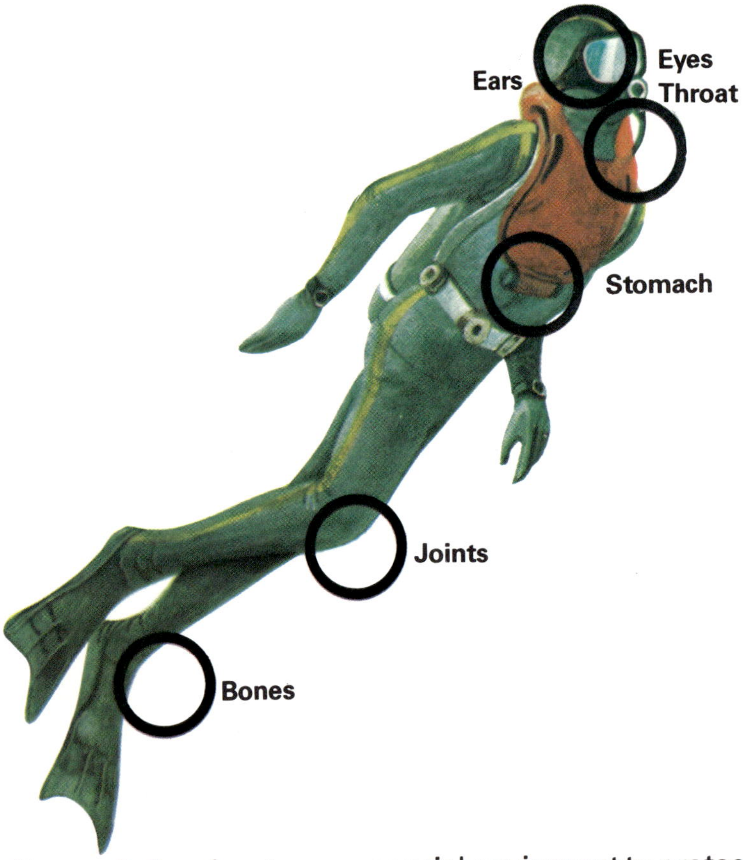

Above: A diver has to wear special equipment to protect himself and help him breathe. The circles above identify the most vulnerable parts of his body.

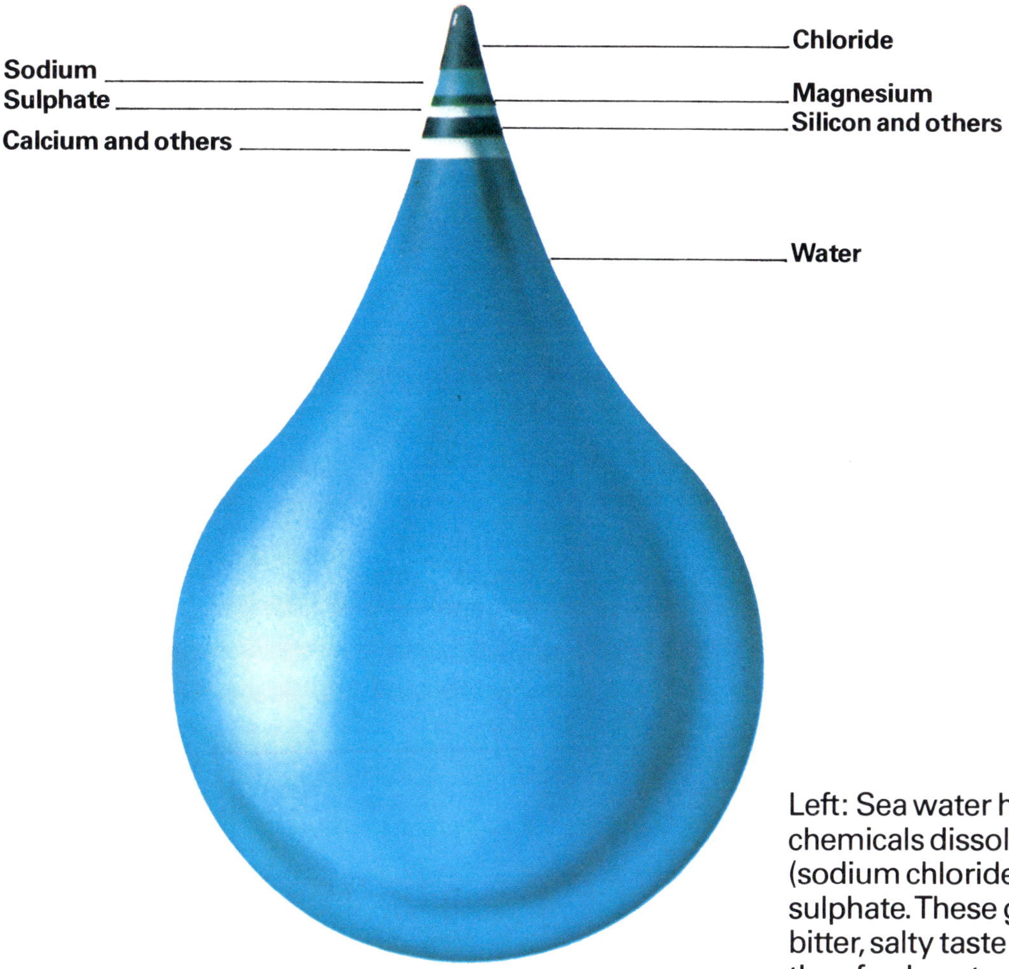

Left: Sea water has many different chemicals dissolved in it, mainly salt (sodium chloride) and magnesium sulphate. These give sea water its bitter, salty taste and make it denser than fresh water.

The human body is designed to live in the earth's atmosphere and to breathe air. Because water is heavier than air, there is more pressure on a man when he is in the sea than when he is on land. Even small changes in pressure affect the way man breathes through his lungs. The pressure of water upon the body and lungs of a diver at a depth of one metre is so great that he can only just breathe through a tube directly connected to the surface. So all divers have to use special breathing equipment in order to stay under water for any length of time. On deep dives they breathe with the help of cylinders of breathing gas. These supply the divers with air at the correct pressure.

The deepest dives any man has made without wearing a diving suit have reached depths of about 150 metres. Under laboratory conditions however, divers have reached much greater depths of up to 609 metres.

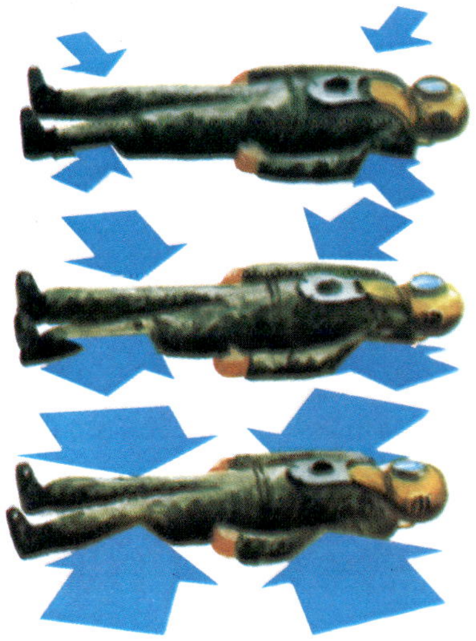

Above: The arrows show the pressure of water on the diver's body. The deeper he goes under water, the greater the pressure.

From early bell to Hard Hat

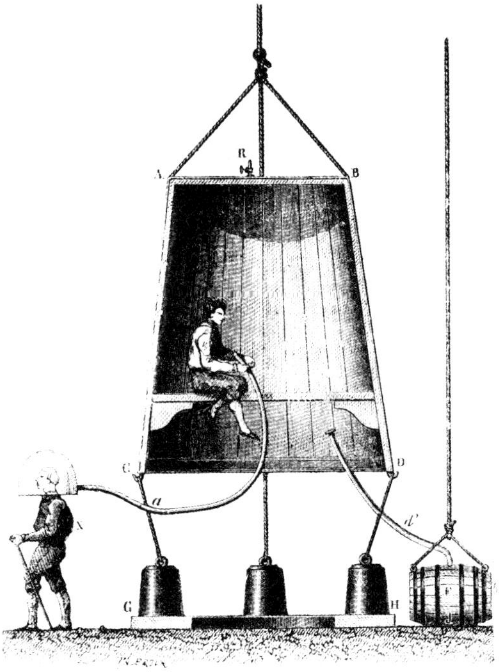

Above: An old drawing of Halley's wooden diving bell. The divers breathed the air through a tube.

For centuries people have dived for sponges, pearls and other treasures without machines to help them. However, it is very difficult for a single swimmer alone to recover treasure and valuable cargo from a sunken ship. He has to work for long periods under water and has to clear away the wreckage before the cargo can be moved. These difficulties led to the design of machines to help divers in their work.

Diving bells

One of the earliest machines to be invented was the diving bell designed in 1690 by Dr Halley, the well-known astronomer. Dr Halley made several experiments before he perfected his design. At first real iron bells were lowered into the sea. As the bell went down, air was naturally trapped inside it. The diver climbed up into the bell to breathe in the air trap. Later diving bells were made of wood. They were kept under water by lead weights. If the diver wanted more air, a weighted barrel of air was lowered into the sea. Air flowed through a tube from the barrel to the bell.

Armoured suits

Halley's diving bell worked, but it had problems. The divers always had to stay close to it and it could not go down very far. All sorts of things were tried to overcome these difficulties. The most successful invention was called the 'Armoured' suit. It allowed the diver to move around much more freely.

The original suit was a simple wooden barrel in which a man was enclosed. There were two tight holes through which his arms went and a window so that he could see out. The barrel was lowered into the sea horizontally so that the man was lying in a swimming position.

Eventually the barrel developed into the suit shown opposite. The head of the diver was enclosed in a steel helmet. He wore a leather suit, but his arms and legs were not covered. This was so that they could move freely. The diver breathed air through tubing which connected him to the surface. But there were still problems with the design. The diver's arms and legs were exposed to the pressure of the water, and this caused pain. It also meant that the diver could not go very deep.

Above: An early 19th century diving suit. It is made of leather and has a steel helmet.

Above: John Deane using his Hard Hat suit to rescue cannon from the wreck of the *Royal George.*

Hard Hat suit

The 'Hard Hat' suit was designed by a man called John Deane. In 1823 he invented a new type of smoke helmet to help men fight fires. Deane then tried the helmet under water. But because it was so heavy, he turned upside down and nearly drowned. The next time he tried the helmet under water, Deane wore lead-soled boots which helped him to stay upright. Since that time lead-soled boots have always been worn with Hard Hat suits. The suits are completely watertight.

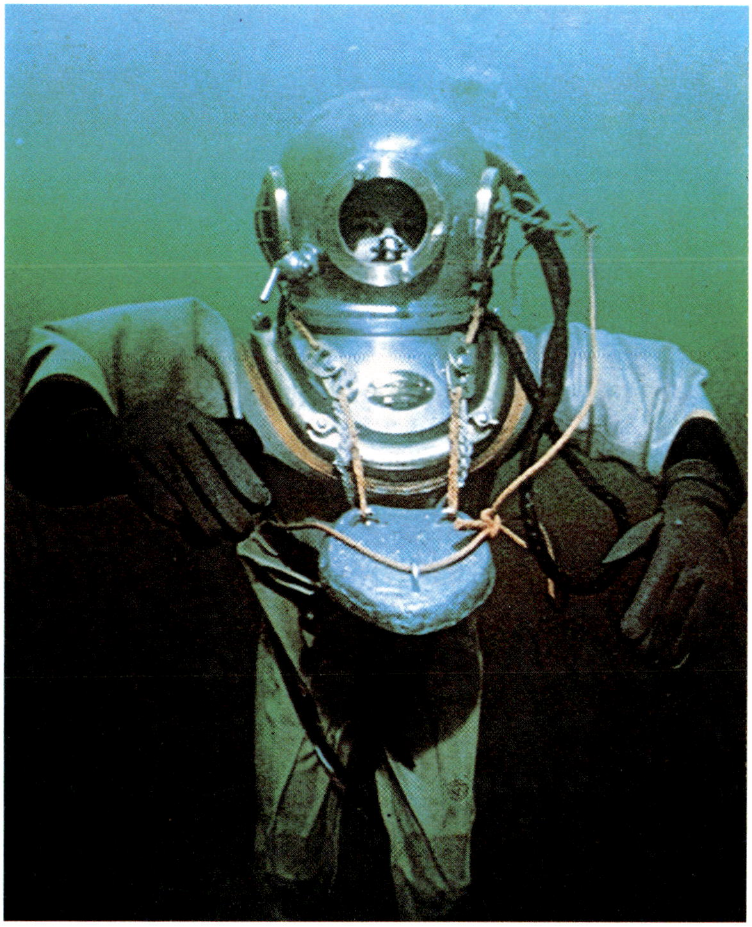

Right: This suit is halfway between the Armoured and the Hard Hat suit. It was built by a German called Kleinert in 1797.

The Aqualung

The Hard Hat diving suit allowed the diver to carry out his salvage work fairly safely. However, he still had to be connected by breathing tubes to the support ship, where the air pump was. This meant that he could not move far from the boat. His suit was also very heavy to keep him under water, and this prevented him from swimming around easily.

During the 1930's scientists tried to find a way for a diver to carry his own air supply. Divers began to carry cylinders of oxygen on their backs. They breathed through tubes connecting the cylinder to a mouth-piece. When the cylinders were empty, the diver had to come up to the surface for a full one. Each cylinder lasted two hours at the most. These early cylinders were mainly used in warfare and even then only by very highly skilled divers.

Leonardo da Vinci's sketch of webbed feet.

Man becomes a fish

The early oxygen cylinders and equipment were very complicated to use. It was also soon discovered that oxygen becomes poisonous to breathe below a depth of twenty metres. But divers cannot use air alone instead. This is because the pressure of the air that the diver breathes has to be equal to the pressure of the water around him.

In the 1940's the well-known French explorer, Jacques-Yves Cousteau, solved the problem. Cousteau and an engineer, Emile Gagnan, invented the self-contained breathing apparatus known as the 'Aqualung' or 'Scuba.' The Aqualung was filled with air not oxygen, but it gave the air to the diver at the correct pressure.

Above and right: The modern diving suit is flexible and can withstand pressure. It also keeps the diver warm. The Aqualung enables the diver to breathe under water. Advances in underwater photography allow him to photograph what he sees.

Left centre: The famous Italian artist, Leonardo da Vinci realised webbed feet make swimming much easier. The flippers divers use today are based on the same principle.

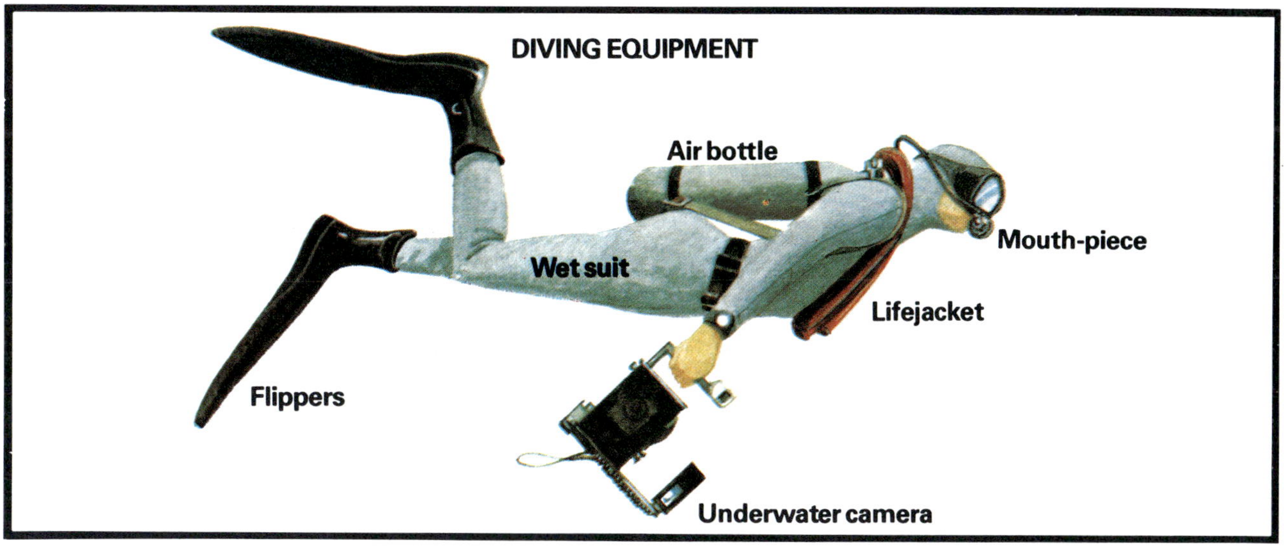

DIVING EQUIPMENT
Air bottle
Mouth-piece
Wet suit
Lifejacket
Flippers
Underwater camera

Deep diving today

In the 1970's men began to look for new sources of oil. The search for new oilfields led to exploration of the seabed in the North Sea. Divers involved in the exploration and the building of oil rigs were required to dive and work in depths of up to 140 metres.

If a diver goes down more than sixty metres, he may suffer from a sickness called the 'bends.' This is because the air he breathes becomes poisonous. To overcome this sickness the diver must breathe a mixture of the gases helium and oxygen. Even with this mixture the diver can only avoid the bends if he comes up to the surface very slowly. A diver working at a depth of 120 metres for one hour would have to take nearly twelve hours to rise to the surface to avoid the bends. Because of this a method called 'saturation diving' was developed which allows divers to stay at their diving depths for up to fourteen days.

Saturation Diving Chambers (SDC) are lowered into the water with the divers inside. The divers come up to rest in the pressurised Deck Decompression Chambers (DDC) on the deck of the support ship and then return to the SDC's. When their work is over, the divers stay in the DDC for a long time, while the pressure slowly returns to normal. In this way divers avoid the bends and can return safely to life on land.

Above: *'Jim,'* an armoured diving suit made in the 1930s. Modern deep diving suits are based on this design.

Three different methods of diving used by modern divers working under the sea. The diver chooses his method according to how deep he wants to go.

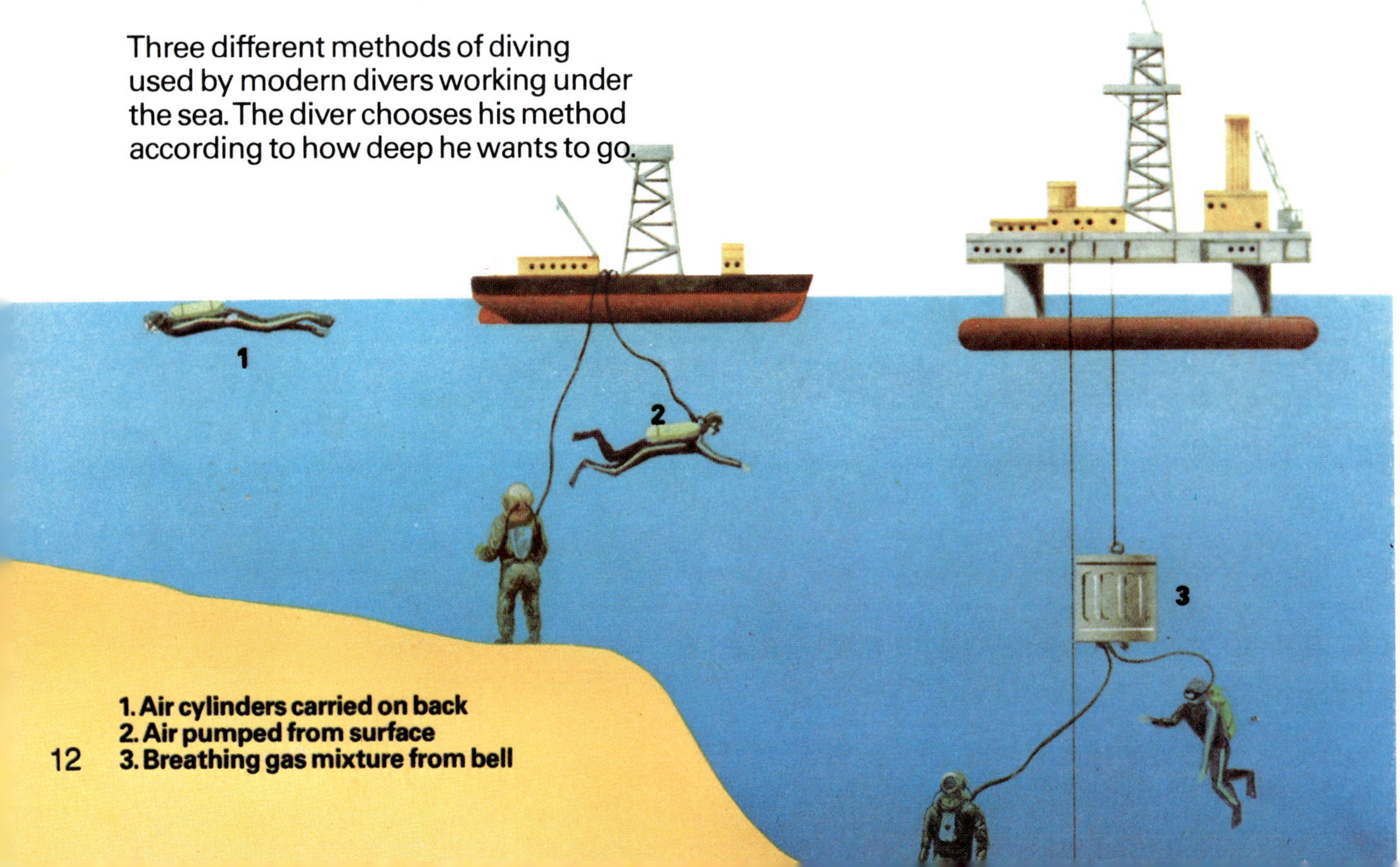

1. Air cylinders carried on back
2. Air pumped from surface
3. Breathing gas mixture from bell

Above: Men enter the SDC. The SDC and DDC are at normal pressure.

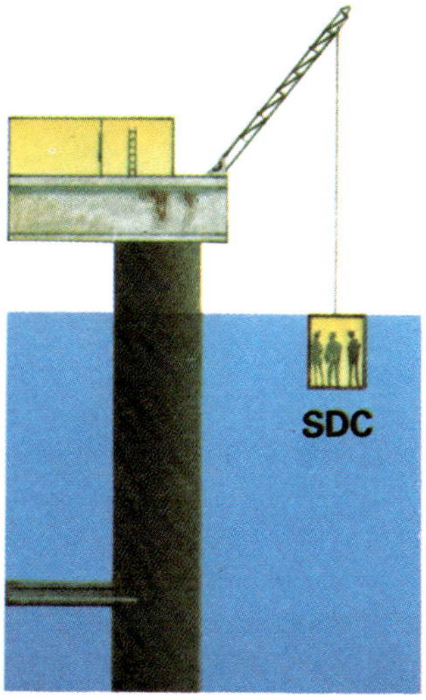

Above: The SDC is lowered and pressurised to the required depth.

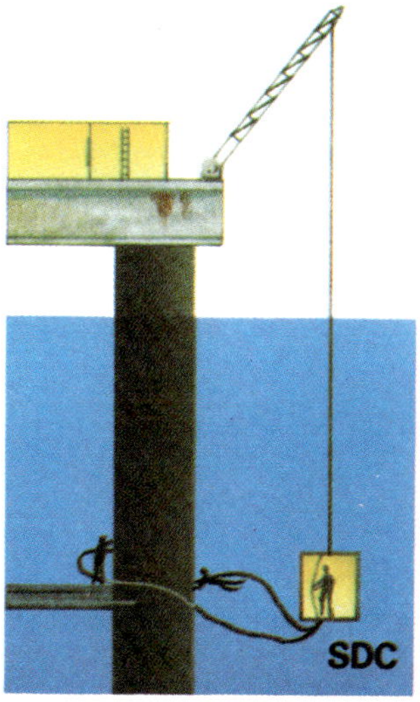

Above: Divers leave the SDC. Breathing gas is piped from the SDC.

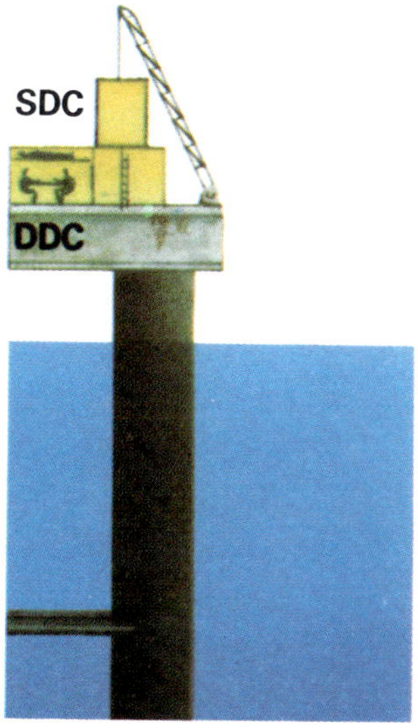

Above: The SDC returns to deck. Divers rest in the DDC at working pressure.

Above: Divers enter a transfer chamber after a long dive.

Homes under the sea

About twenty years ago, scientists began to study what happens when divers stay under water for a long time in specially built houses or habitats. They thought that many of the difficulties associated with deep sea diving could be solved if the diver worked, ate and slept at the same depth. In the middle of the 1960's therefore, several countries built habitats. The habitat *Conshelf 3* was built for Cousteau and, in 1965, divers stayed in it for three weeks at a depth of 128 metres. The Sealab experiments carried out by the United States Navy were abandoned in 1969.

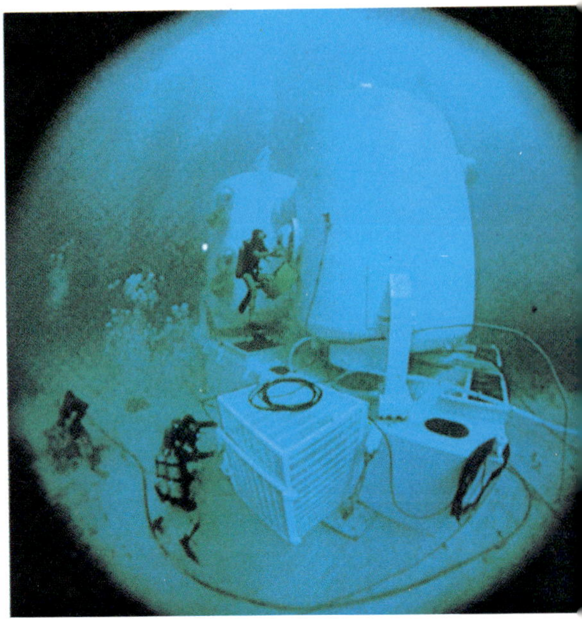

Above: This is a view of *Tektite* in position on the seabed with divers working outside it.

Left: Special submersibles were built for experiments as part of the underwater habitat programmes. They were used to do things like record and measure the effects of sound under water. Some also studied the problems of pressure and light under water.

Problems in living under the sea

The main difficulties experienced in underwater habitats are the usual ones of the pressure of the water and the cold. Light is a problem when the divers explore in the dark sea outside the ship. It is also necessary to have a large and costly support ship nearby in case anything goes wrong. Most of the experiments in the 1960's were too expensive and did not solve many of these difficulties.

One of the few habitats to survive into the 1970's was the American *Tektite*. This was placed in a depth of fifteen metres of water off the Virgin Islands in the West Indies. The reason for its success was that the shallow water,where it was, was warm and clear. The usual problems just did not arise.

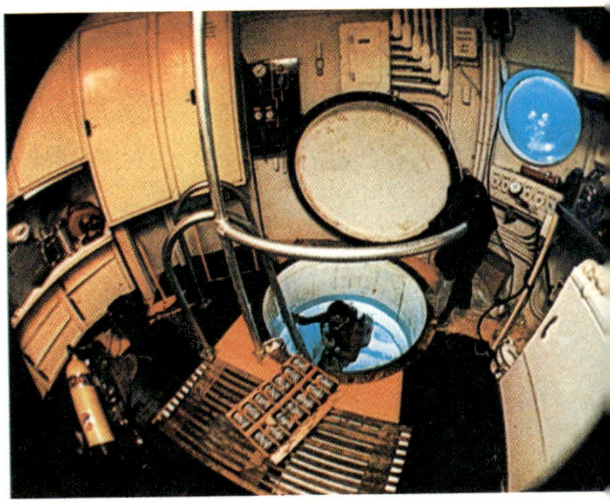

Above: The space inside a habitat is filled with instruments. Divers have to enter by a special chamber.

Above and left: A Sealab habitat before launching and (left) on the seabed. Sealab carried out experiments under water just as Skylab did in space. It was one of the most important habitat experiments of the 1960s.

Submarines and war

One of the main reasons why underwater technology has developed to such a high level is because the sea is very important during war. Governments are prepared to spend a lot of money to improve undersea weapons and submarines.

One of the most successful early submarines was called Bushnell's *Turtle*. It was used by the Americans as far back as 1775 in their War of Independence. The first torpedo, the Whitehead Torpedo, was invented in 1868. It could be launched under water from a submarine and therefore allowed submarines to keep their distance during an attack.

In 1920, however, men learnt to detect the presence of submarines by using sound waves. The waves were transmitted through the water and bounced back if they hit anything. Submarines themselves were now hunted.

Above: Bushnell's *Turtle* was used to attack the British fleet in 1775. It was made of wood.

Above: *H.M.S. Conqueror* is a Fleet submarine used to hunt Polaris submarines.

Above: Polaris submarines fire their missiles under water from tubes like these.

Nuclear submarines

Modern submarines are powered by nuclear reactors. They do not have to surface to recharge their batteries like the early submarines. Some of them carry Polaris missiles. These atomic missiles are launched under water and can reach targets thousands of miles away. Other nuclear submarines, called 'Fleet' submarines, are used to hunt the Polaris submarines. These modern submarines are the most important part of any navy.

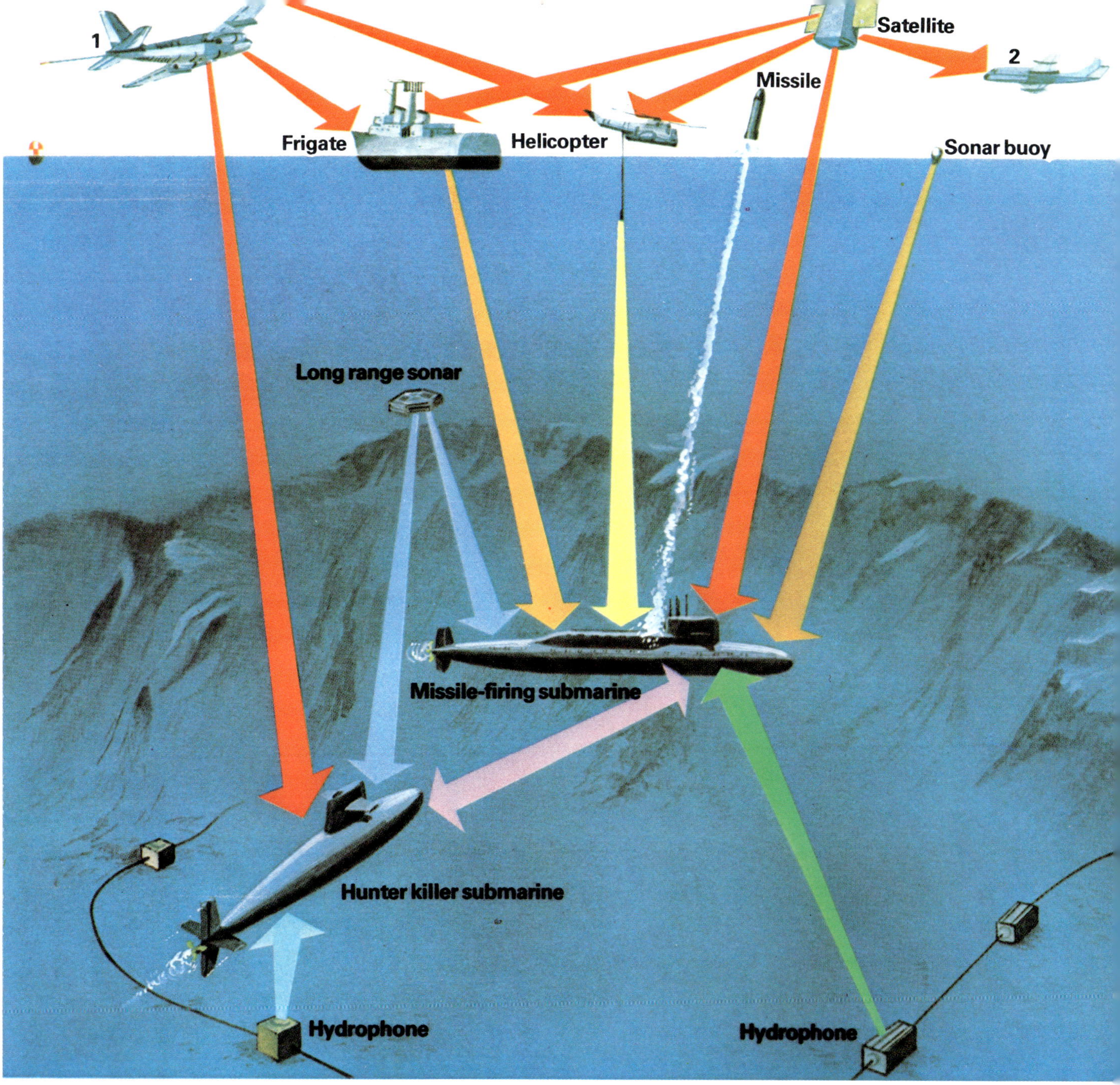

The atomic missiles that modern nuclear submarines carry are very dangerous. Because of this, new methods have been developed to detect the submarines in the middle of the ocean. They can then be destroyed before they fire their missiles.

The submarines are traced by using 'sonar' devices. These devices can be placed on the seabed, carried by surface ships or even used from a helicopter or aircraft. They detect the submarine by using sound waves. Fleet submarines use sonar devices to track enemy submarines before destroying them with torpedoes. The most modern navies also have satellites which can detect the movement of submerged submarines.

Above: In order to find and destroy missile-carrying submarines, surface ships, aircraft and other submarines help each other to search the sea. They do this by using 'sonar' devices. All the ships and aircraft talk to each other by means of a satellite. The hunted submarine tries to keep as quiet as possible to avoid detection. The arrows above represent the paths of communication between the searching craft.

1. Viking aircraft. 2. Nimrod anti-submarine patrol.

To the bottom of the world

The First World War was the first war in which submarines played an important part. They could have been put to more use however if they had been able to dive deeper. After the war, scientists began to do more research into deep diving. They used round craft called 'bathyspheres' for their experiments.

Bathyspheres are the best shape for resisting the large pressures experienced in great depths of water. They were made of steel and hung from a support ship. On August 15th 1934 an American bathysphere reached a depth of 924 metres. This was a world record at the time. The sphere was one and a half metres in diameter and had steel walls four centimetres thick.

Above: Much of the limited space inside a bathysphere was taken up by equipment.

Above: This French machine was an early attempt to build a bathysphere that could move about under water.

Most bathyspheres at this time were lowered into the water at the end of a wire rope. The people in the sphere had to trust the strength of the wire. If it broke, the sphere would sink to the bottom of the sea because it was so heavy. Escape from this depth would have been impossible.

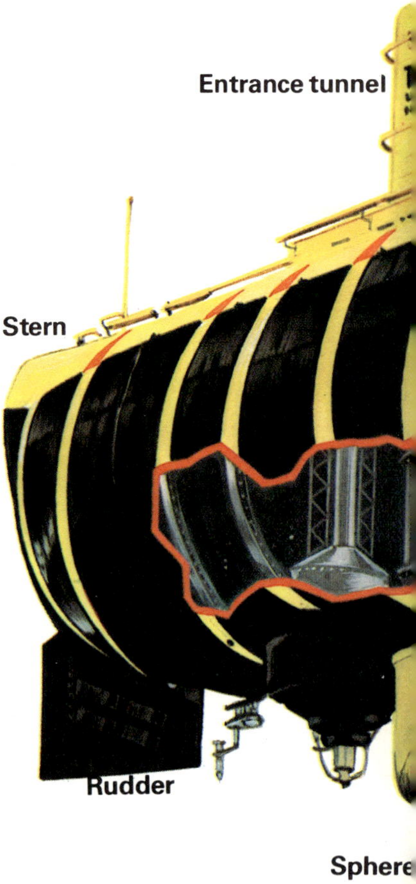

The bathysphere hung from a support vessel, so it could not move about and investigate anything. This was a great disadvantage. To overcome this a different version was made. It was called the 'bathyscaphe' and was developed by Professor Auguste Piccard.

The bathyscaphe consisted of a sphere which hung from a large buoyancy tank. The tank was filled with petrol and iron pellets. When the bathyscaphe was ready to dive, petrol was pumped from the tank and water took its place. Water is heavier than petrol, so the craft sank. When the iron pellets were released from the tank, the craft floated to the surface. On 23 January 1960, the bathyscaphe *Trieste* reached a depth of 11,000 metres.

Above: Later developments of the bathyscaphe led to the design of *Trieste II.*

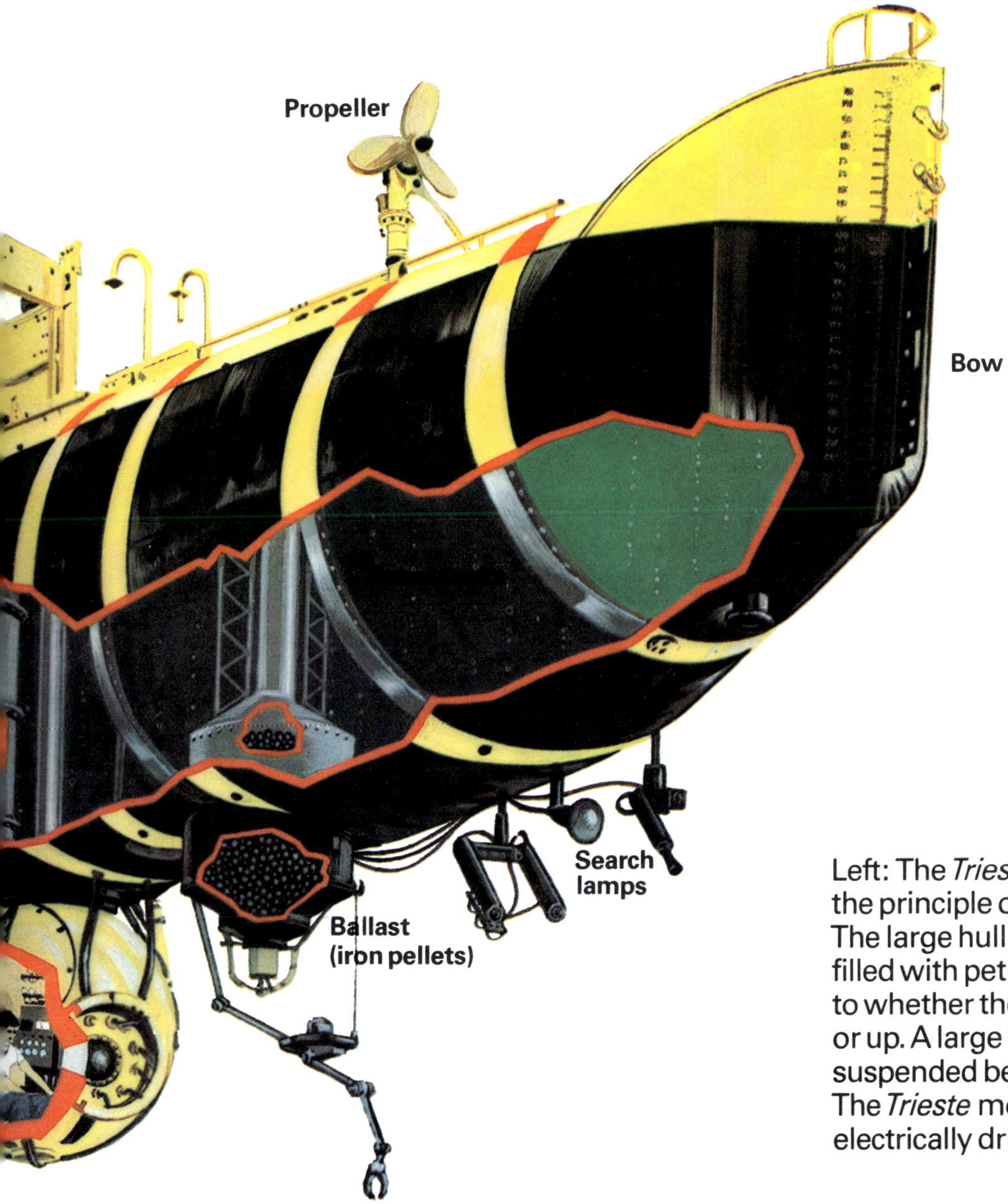

Left: The *Trieste* was designed on the principle of the bathyscaphe. The large hull was like a tank. It was filled with petrol or water according to whether the craft was going down or up. A large bathysphere was suspended beneath the craft. The *Trieste* moved around by using electrically driven propellers.

Working craft of today

Nowadays craft used commercially for underwater work and undersea exploration are called submersibles rather than submarines. Submersibles are smaller and are not designed to stay under water as long as submarines. They are normally carried to the diving area by a mother ship. Most of the work they are used for is connected with undersea engineering.

Some submersibles have specially designed claws which can be operated from inside the craft. They can carry, lift and perform a number of difficult tasks without using divers. Other submersibles have a lockout chamber. The divers enter and leave the craft through the chamber. This type is used on jobs, like welding, where manual skill is required.

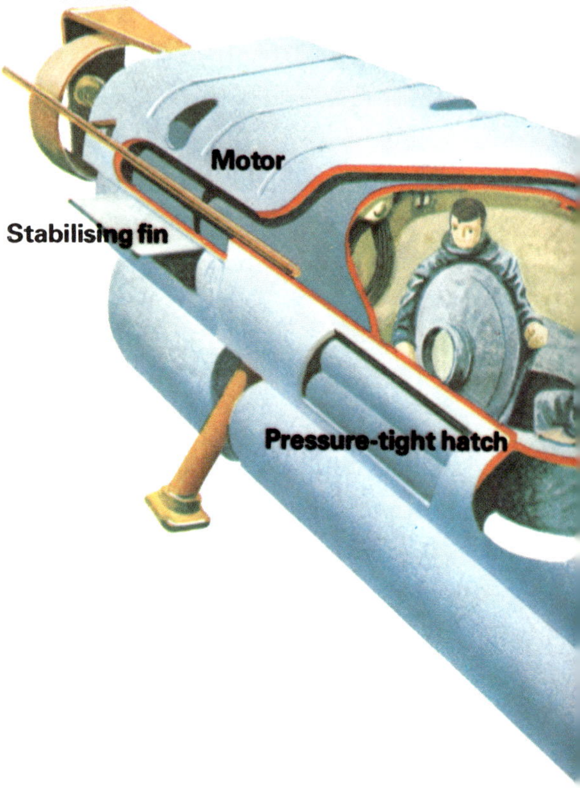

Above: *Consub I* is an unmanned submersible used for research of the seabed.

Not all submarines are used for engineering work. Some have been built to carry out special research. This research is mainly concerned with studying the oceans and the fish population as well as charting the seabed. Some of the submersibles used for this sort of work are purely automatic. They are fitted with television cameras and are controlled from the mother ship. For some of the simpler types of research, submersibles are towed behind a surface ship. They use special instruments to record things like water temperature and light.

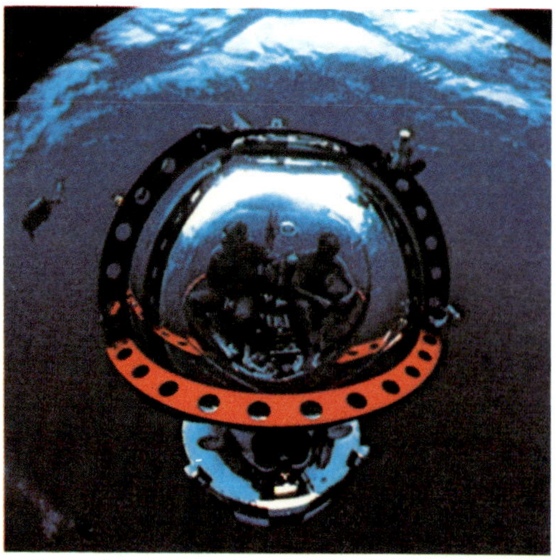

Above: This submersible was built by the U.S. Navy as an experimental observatory.

Below and right: A lockout submersible. When a diver leaves the craft, he is locked into a lockout chamber. Water is allowed to come in. When the water pressure inside the chamber is equal to the pressure of the sea outside, the door opens and the diver swims out. When he goes back, he is locked in again and the water is pumped out.

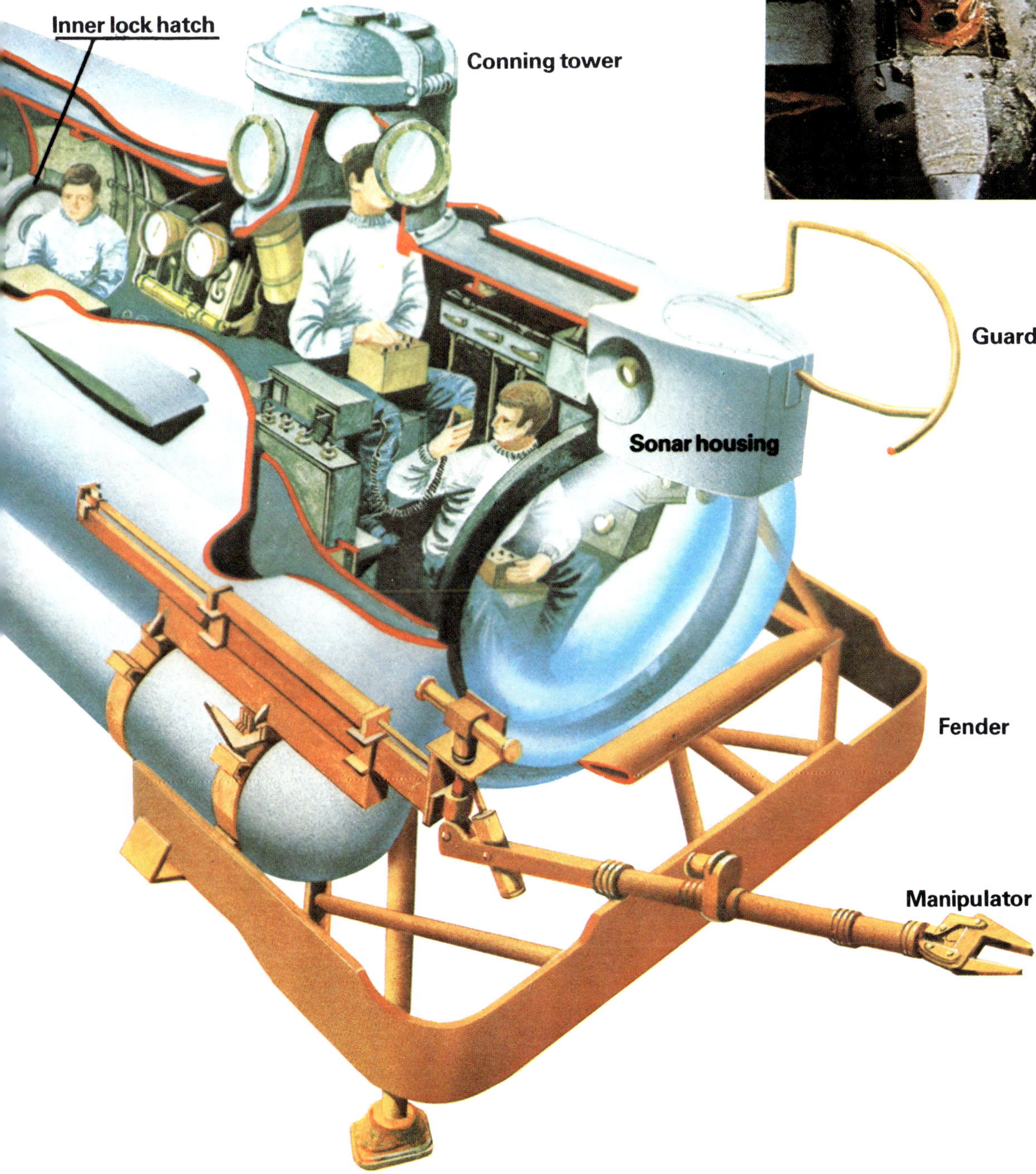

Food

The sea has always been a source of food for man. Today over-fishing has become a problem. Over-fishing occurs when so many fish and animals are taken from the sea that some species are threatened with extinction. This has happened with the cod and other common species. Now countries throughout the world have agreed to catch only a certain number of these fish each year.

Above: Fishermen examining their catch. Fish are an important source of protein in our diet.

Fish farming

A lot of underwater research has been concerned with the study of fish. We now know a great deal about the habits of certain species. This knowledge has been used to set up fish farms. The fish are kept in cages in lakes where they are fed and looked after so that they do not become ill. They breed and produce more fish. When they are large enough the fish are taken out of the lake for eating. Oysters, lobsters and crabs are also farmed in this way. In some parts of the world seaweed is harvested for sale both as a food and a fertiliser.

New sources of food

At present most fishing for food is carried out in quite shallow water. But because there are more people in the world today we need more food. This means we will have to go fishing in deeper ocean waters to find new species of fish. This sort of research is an extremely important part of current undersea exploration.

Below: Three modern methods of catching fish. The nets are dragged behind trawlers.
1. Drift netting
2. Seine-purse netting
3. Trawling

1

2

3

Fishing today

The majority of fish eaten today are caught by trawlers. Many of these boats have huge freezers on board. The fish are processed and frozen while the trawler is at sea. The largest trawlers can carry as much as 700 tonnes of fish.

Some boats use a special device to help them detect large shoals of fish. The device is rather like the sound-ranging system that is used to track enemy submarines. The shoal shows up on a screen fitted on the bridge of the trawler. When the skipper sees this he can head the boat towards the fish.

Most trawlers fish in fleets which consist of a mother ship and several small trawlers. The mother ship processes and stores the fish. Large boats are sent out from the harbour to collect the fish for marketing. Other ships are sent out to refuel the trawlers at sea so that the fleet does not have to return to shore until the fishing season is over.

Above: Mussels are now grown successfully on fish farms.

Below: Modern trawlers use sound wave techniques to locate shoals.

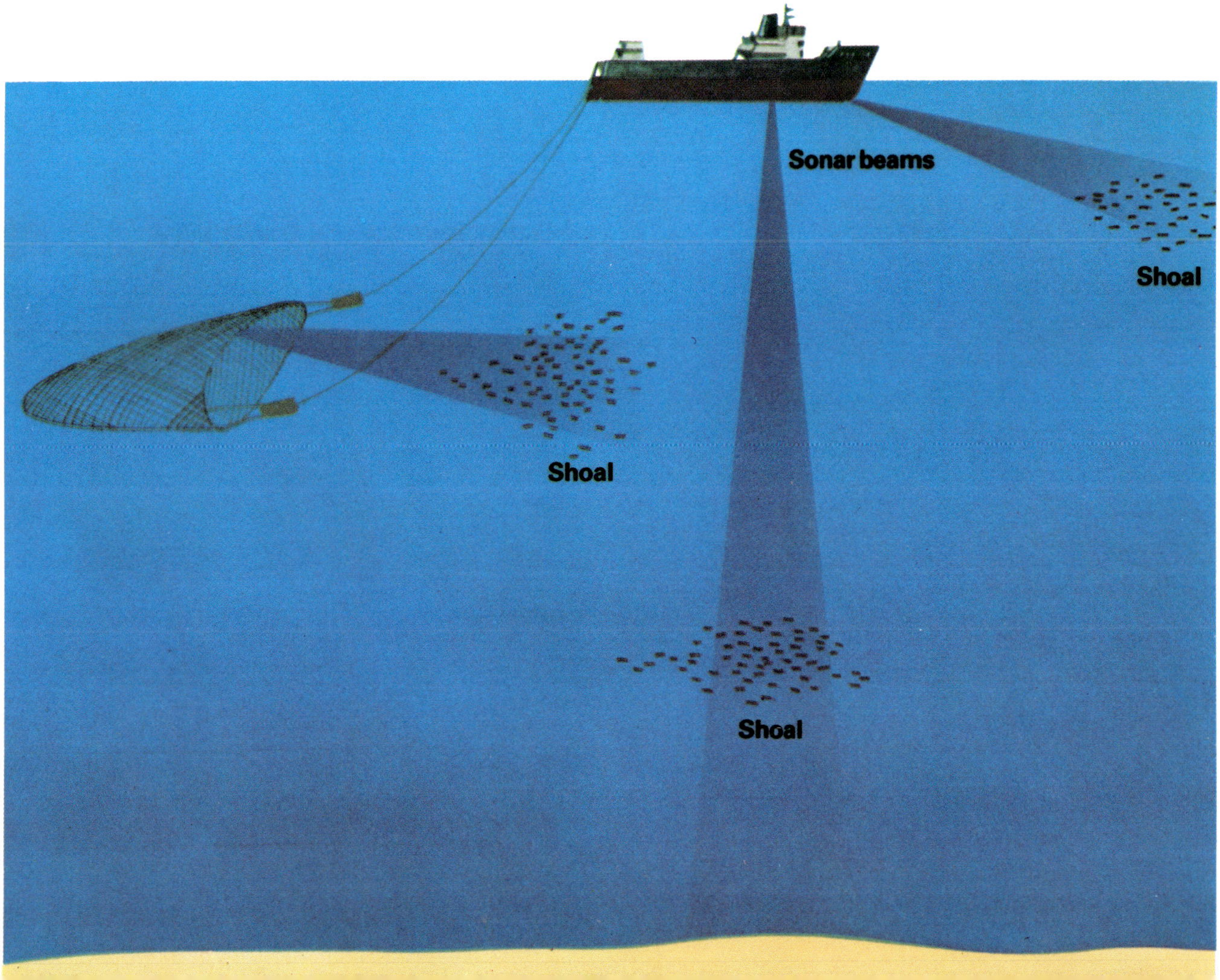

Minerals and energy

Most of the machines we use today run on some kind of fuel such as oil or natural gas. Oil is converted into petrol for our cars. Natural gas is piped into our homes for heating and cooking. Many oil and natural gas fields have been found under the sea. A great deal of money is now being spent to explore new areas such as the North Sea and the Bass Straits of Australia. Even if this exploration is successful, however, there are still many problems in bringing the fuel to the surface.

To build an oil well, you have to drill a hole in the seabed. This is often difficult because most oilfields are found under very deep water. The oil from the well is pumped through pipes to the production platform on the surface of the sea. The pipes need constant inspection to avoid leakages. The impurities are taken out on the production platform before the oil is put into tankers or piped to a refinery. Finally at the refinery it is made into petrol or diesel oil. The whole process is very expensive and dangerous.

Above: Salt pans in Morocco.

Below: Scientists working on the 'Nodding Duck' experiment. It harnesses basic sea energy to generate electricity.

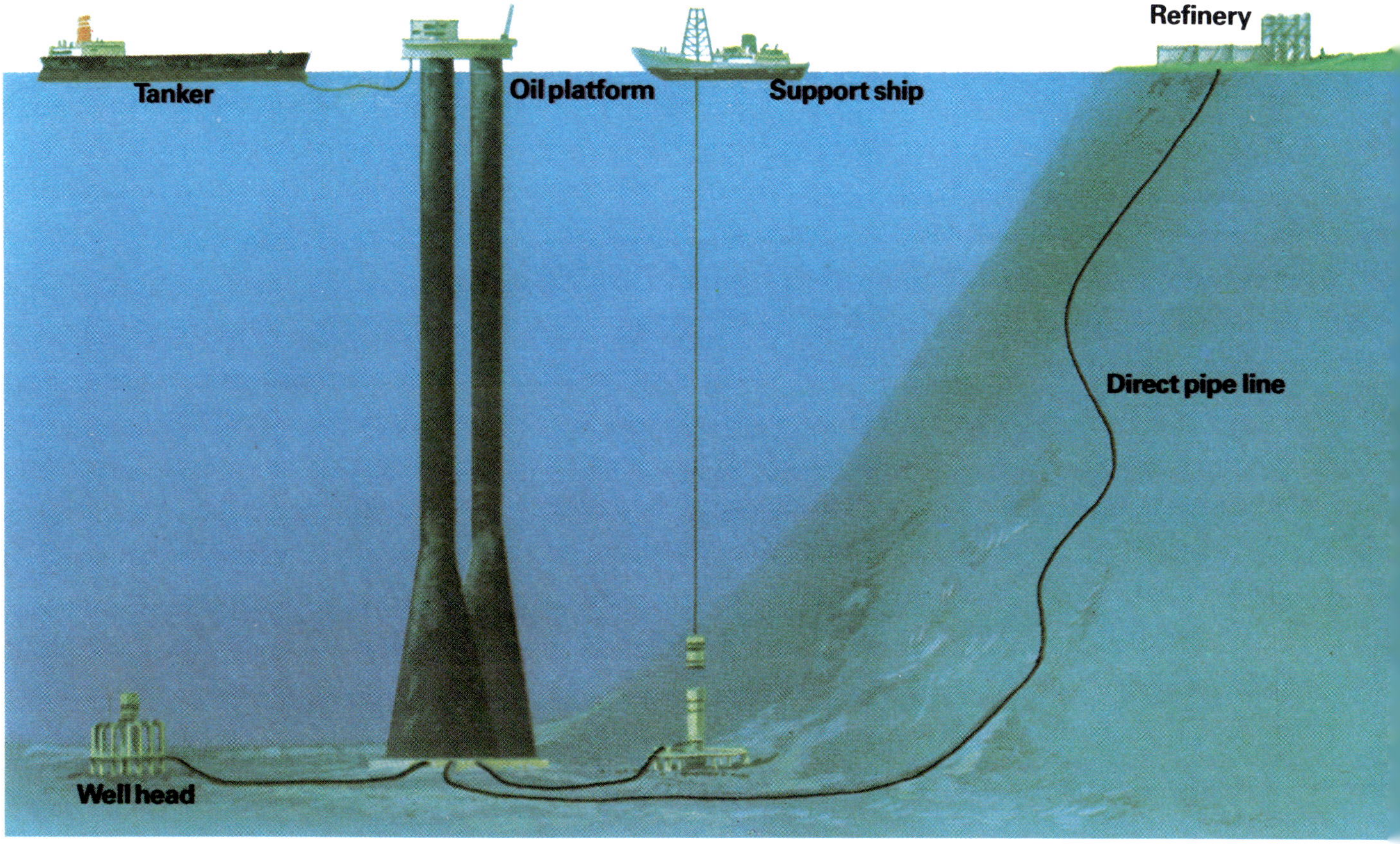

Above: Oil is usually pumped up to an oil platform. It is only piped direct to the refinery if the coast is close.

Above: Huge quantities of manganese nodules are dredged up from many parts of the seabed.

Alternative sources of energy

The sea is a source of energy in its own right. This energy shows itself in waves that break on the seashore and in the rise and fall of tides. Energy also exists where there is a temperature difference in the sea. It is often much colder at the bottom of the sea than at the top. Because there is a flow of heat, there is a flow of energy. This flow of energy can be used to drive turbines which produce electricity.

This basic sea energy is not like gas and oil which are burnt to release energy. Therefore we either have to adapt our machines so that they can use it, or design special new machines.

One such machine is called a 'Nodding Duck'. It has a special float which is moved backwards and forwards by the waves. This nodding movement is then used to turn an electricity generator.

Minerals

The sea is also a very important source of minerals, like salt and manganese. Sea water has been used to make salt for centuries. The water is channelled into salt pans where it evaporates leaving behind salt crystals.

Exploring for fun

Nowadays many people explore the sea as a hobby. There are many different ways of doing this. In some parts of the world, such as the Red Sea and around the coast of Japan, special viewing towers have been built. People enter the towers above the waterline. Then they go down the tower to look at the fish and plant life outside through thick glass viewing windows.

Other people learn to dive. They use either a snorkel or an aqualung to help them breathe. Some divers enjoy simply swimming around under water. Others do more exciting things like spear fishing or underwater photography. In some places there are even underwater parks for the divers to visit. The attractively coloured fish in these parks are protected by law.

Occasionally underwater divers find wrecks on the seabed. If the wreck is an old one it may be important because it can tell us what life was like on ships in days gone by. Archaeologists try to recover and preserve as many of the articles from the wreck as possible. Sometimes the wreck itself is lifted up.

Above: An observation tower for viewing underwater life.

Right: A diver exploring an underwater park in America.

Below: An archaeological site in the Mediterranean Sea.

Over-exploitation

Man takes the sea for granted. He often uses it too much without realising that he can destroy the delicate balance of nature. Over-fishing, for instance, upsets the natural breeding habits of certain species. We must take care not to over-exploit the sea's resources.

Above: A guillemot covered in oil. He has to be washed in special solvents to be cleaned.

Pollution

We also use the sea as a great big rubbish dump for everything, and this causes pollution. Chemicals used in agriculture are washed into the rivers by rain. Rivers drain into the sea and so the sea becomes polluted. Waste from factories also pollutes water. Another big danger comes from the nuclear waste we put in the sea. If it leaked from its containers all life in the sea would suffer. In some enclosed seas, such as the Great Lakes of North America, pollution is so bad that even the fish become poisonous.

Oil pollution

Oil is transported round the world in huge tankers. When one of these tankers collides with another or is wrecked, enormous amounts of oil escape and cause pollution. The oil slicks do not sink because oil is lighter than water. Any bird which gets covered in the oil will die unless it is given help quickly.

Above: Sewage and waste emptied into the sea cause pollution.

Special traffic routes have been laid out in busy shipping areas to reduce the number of large tanker accidents. When these accidents do happen special detergents are sprayed on the oil slicks by boats and planes. The detergents break the oil slicks up and make them heavier so that they sink. Some boats have great vacuum cleaners which suck up the oil from the surface of the sea.

World wide control of pollution

The dangers of pollution can only be overcome when all the countries of the world agree to control it. Whatever happens to a piece of water belonging to one country eventually affects another country. This is because of the strong ocean currents. They flow around the world and carry pollution from continent to continent.

Right: The tanker *Amoco Cadiz* ran aground because her steering broke.

Future undersea exploration

Undersea exploration is both difficult and expensive. At present it is not worthwhile building underwater habitats for frogmen to live in. There always has to be a ship standing by in case of danger. There are also still many problems connected with living under water for a long time.

Despite these difficulties, undersea exploration must go on. We need more energy to use as fuel and we need more food to feed our growing population. The sea can supply many of these needs. If we can harness the basic energies of the sea we can generate electricity. If we learn more about the natural habits of fish we can farm fish more efficiently.

However, if we are to get the most out of the sea in years to come, we must also be careful to use it properly today. We will have to guard against over-exploitation and pollution. Man's future on earth may depend on the riches of the sea.

Above: A special submersible built by the United States Navy which can dive down and rescue the crew from a sunken submarine.

Above: This ship carried out an experiment called 'Operation Mohole'. It drilled the deepest hole in the world to find out more about the earth's crust.

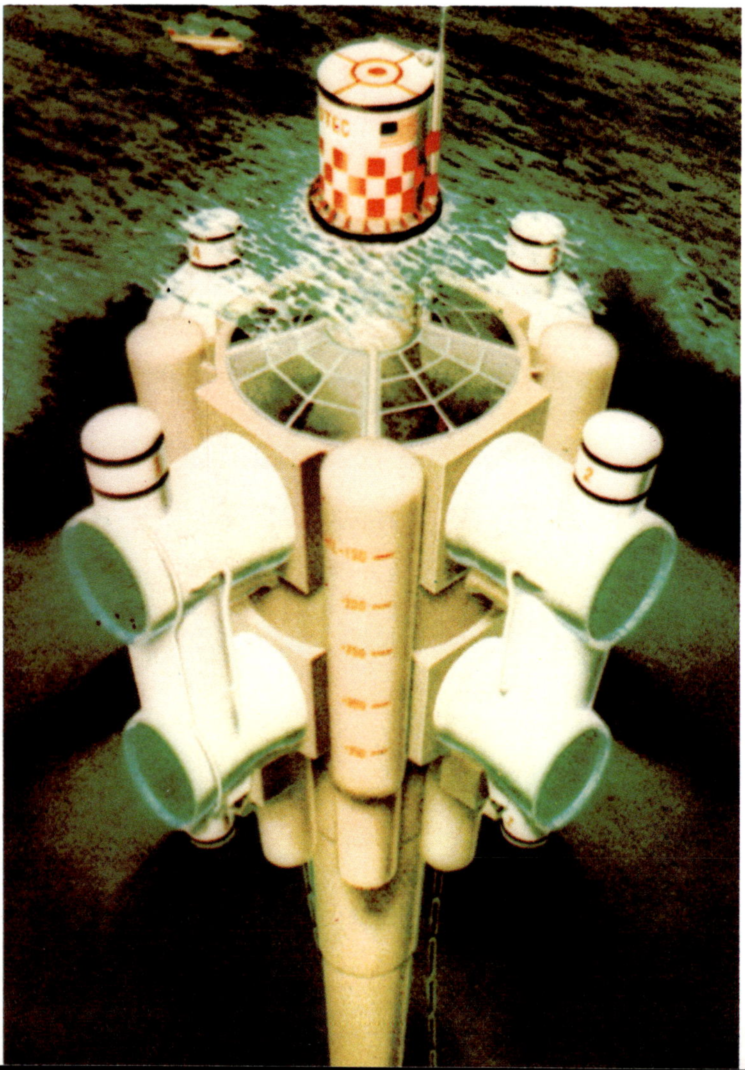

Left: *Otec* has been designed to produce energy from the sea. It uses temperature differences in the water to drive turbines which generate electricity.

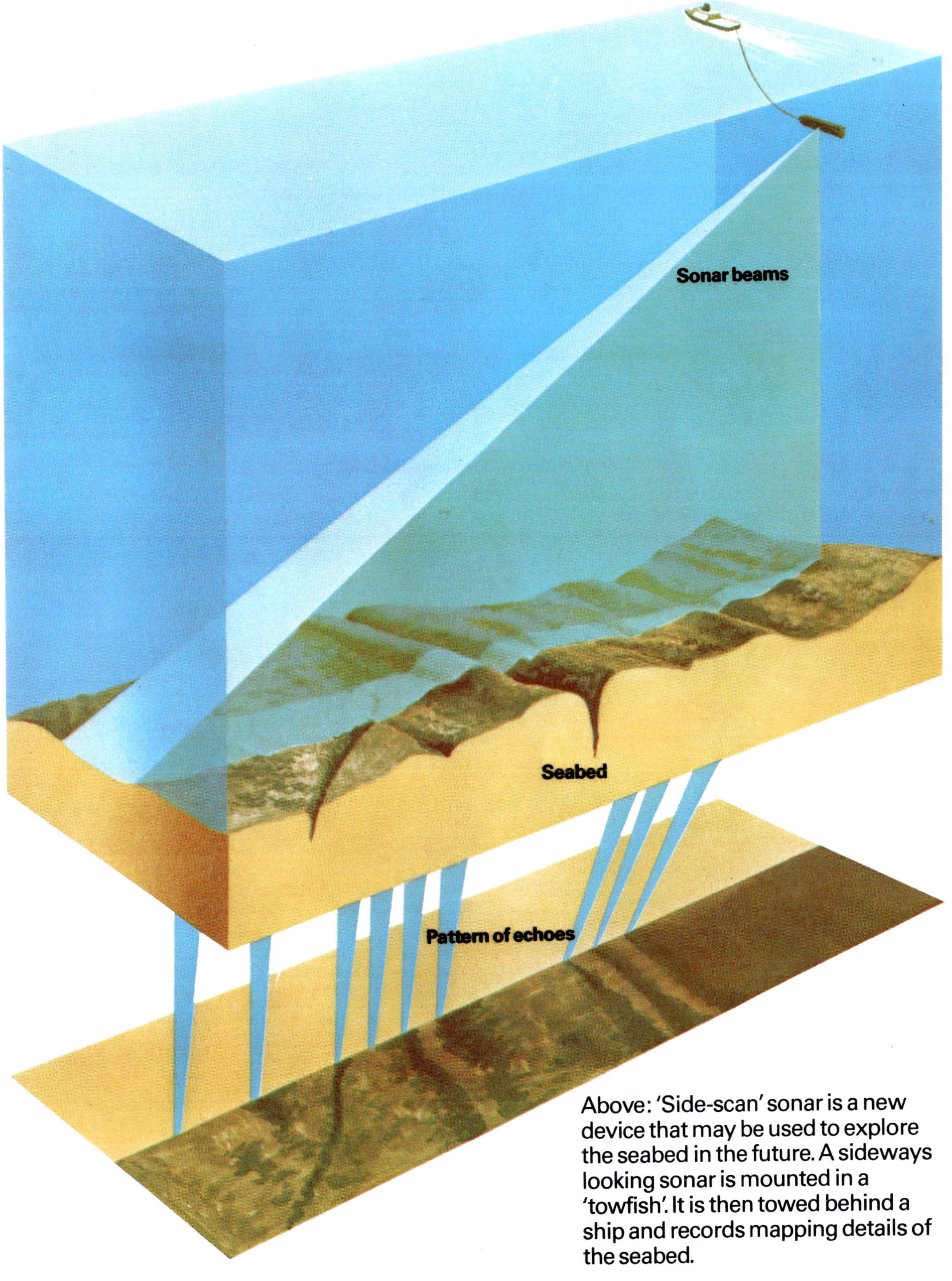

Above: 'Side-scan' sonar is a new device that may be used to explore the seabed in the future. A sideways looking sonar is mounted in a 'towfish'. It is then towed behind a ship and records mapping details of the seabed.

Memorable dives

1935. The armoured suit called *Jim* is used to dive 150 metres to find the wreck of the *Lusitania* off the coast of Southern Ireland.

1969. A submersible called the *Ben Franklin* drifts in the Gulf Stream for a month at depths varying between 180 and 600 metres.

1934. Beebe's bathysphere reaches a depth of 924 metres. This a world record at the time.

1960. The bathyscaphe *Trieste* establishes a new world record by diving to a depth of 11,000 metres.

Index

Surplus gas
Production platform
Drilling derrick
Telecommunications
THISTLE
211/18

Semi-submersible rig

1 2 3 4 5 6 7 8—U—88 87 86 85 84 83 82 81